ANNALES ACADEMIÆ SCIENTIARUM FENNICÆ

MATHEMATICA

DISSERTATIONES

133

MEROMORPHIC SOLUTIONS OF SOME COMPOSITE FUNCTIONAL EQUATIONS

HELI SILVENNOINEN

University of Joensuu, Department of Mathematics

To be presented, with the permission of the Faculty of Science of the University of Joensuu, for public criticism in Auditorium M4 of the University, on June 4th, 2003, at 12 o'clock noon.

HELSINKI 2003
SUOMALAINEN TIEDEAKATEMIA

ISSN 1239-6303
ISBN 951-41-0928-7

Received 30 April 2003

2000 Mathematics Subject Classification:
Primary 30D05; Secondary 39B32, 30D35.

YLIOPISTOPAINO
HELSINKI 2003

Acknowledgements

I would like to express my sincerest gratitude to my advisor, Professor Ilpo Laine, for his valuable advice and constant encouragement. His experienced guidance had a great impact on my work.

I am grateful to Professors Gary Gundersen and Kazuya Tohge for reading the manuscript and making many valuable comments and suggestions that greatly improved the final version.

My warm thanks are due to the faculty and the staff at the Department of Mathematics of the University of Joensuu for providing me a friendly environment and for all the help they offered me.

For the financial support I am indepted to the Väisälä Foundation, the University of Joensuu, the Graduate School on Mathematical Analysis and Logic and the Magnus Ehrnrooth Foundation.

Finally, I want to thank my family and all my friends for their support and encouragement.

Joensuu, April 2003 Heli Silvennoinen

Contents

Introduction

Functional equations have been widely studied in the real domain. However, in the complex plane functional equations have attracted considerable attention during the last few decades only. In the 1970's R. Goldstein wrote a series of articles concerning the value distribution of solutions of certain functional equations containing composite functions. A substantial part of his research is concerned with functional equations of the form $f(g(z)) = q(f(z))$ and $f(g(z)) = q(f(z)) + h(z)$, where f and h are meromorphic functions, g is entire and q is a rational function of order $k \geq 1$. Goldstein applied the Nevanlinna theory to prove several results on the growth of the functions involved. See, for example, [8], [9], [10] and [11]. During the last decade meromorphic solutions of functional equations have been studied, for example, by W. Bergweiler et al., see [4], [5] and by J. Heittokangas et al., see [14].

J. Ritt [20] wrote a classical paper on the Poincaré equation

$$f(cz) = R(f(z)),$$

where $c \in \mathbb{C}$, $c \neq 0, 1$, and $R(f)$ is a rational function in f. In a collection of research problems [21], L. Rubel posed the following question: What happens if $R(f)$ is replaced by $R(z, f)$, where $R(z, f)$ is rational in both variables? K. Ishizaki [16] considered the equation

$$f(cz) = a(z)f(z) + b(z),$$

where $a(z)$ and $b(z)$ are meromorphic functions. He showed that under certain conditions on the coefficients, that equation possesses at least one meromorphic solution and especially if the coefficients are rational, then all transcendental meromorphic solutions are hypertranscendental. Earlier, H. Wittich [23] studied the same equation in the case when $a(z)$ and $b(z)$ are polynomials. Recently, G. Gundersen et al. considered meromorphic solutions of the general equation

$$f(cz) = R(z, f(z)),$$

see [12], proving several results on the growth and value distribution of solutions. Moreover, they offered a detailed analysis on the number of distinct meromorphic solutions in the special case of $f(cz) = A(z) + \gamma f(z) + \delta f(z)^2$, where $c, \gamma, \delta \in \mathbb{C}$, $|c| > 1$, $\delta \neq 0$ and $A(z)$ is entire.

In this paper we study the growth and existence of meromorphic solutions of functional equations of the generalized form

$$f(p(z)) = R(z, f(z)), \tag{1}$$

where $p(z)$ is a polynomial and the coefficients of $R(z, f)$ are small functions as compared with f. The problem is that explicit transcendental meromorphic solutions to these equations are hard to find. Since we are interested in meromorphic solutions in the whole complex plane we can use the Nevanlinna theory as a tool to study the properties of solutions.

We begin by recalling basic notations of the Nevanlinna theory and by introducing some important results about composite functions in Section 1. Section 2 is dedicated to studying the equation

$$f(az + b) = R(z, f). \tag{2}$$

We prove for example that any transcendental meromorphic solution of (2) is of regular growth. However, our main interest is devoted to the case when $\deg p \geq 2$ in (1). In Section 3, we start to consider the equation (1) in its general form. Most of the results in this section concern the growth of the meromorphic solutions in terms of the characteristic function. In Section 4, we consider growth relations between the solution f and the coefficient a in the equation

$$f(p(z)) = a(z)R(z, f(z)). \tag{3}$$

Section 5 is devoted to considering a special case of (3), that is, the equation $f(p(z)) = a(z)f(z)^m$. We are able to prove results concerning the existence of meromorphic solutions. A connection between the meromorphic solutions of the equation

$$f(p(z)) = a_0(z) + a_1(z)f(z) + \cdots + a_{m-2}(z)f(z)^{m-2} + f(z)^m$$

and of a certain Riccati differential equation is considered in Section 6. Finally, in Section 7 we study meromorphic solutions of equations of the form

$$f(p(z)) = a_0(z) + a_1(z)f(z) + \cdots + a_m(z)f(z)^m \tag{4}$$

under restriction on the growth, and respectively, the number of poles, of the coefficients in (4).

1. Notations and background

In this section we recall some definitions and preliminary results from the Nevanlinna theory. We also introduce some useful lemmas about the characteristic function of the composite functions. These lemmas will be used in several occasions later on. We note that by a meromorphic function, we always mean a function meromorphic in the whole complex plane.

1.1. Nevanlinna theory

We use the standard notations of the Nevanlinna theory, i.e. $m(r,f)$, $n(r,f)$, $N(r,f)$ and $T(r,f)$ denote the *proximity function*, the *non-integrated counting function*, the *counting function* and the *characteristic function* of f, respectively. For the integrated counting function for distinct poles of f and for the integrated counting function for multiple points of f we use the notations $\overline{N}(r,f)$ and $N_1(r,f)$, respectively. Furthermore, the *order of growth* and the *lower order* of a meromorphic function f are denoted by $\rho(f)$ and $\mu(f)$, respectively. When $\rho(f) = \mu(f)$ we say that function f is of *regular growth.* Moreover, we say that a meromorphic function g is *small* with respect to f if $T(r,g) = S(r,f)$, where $S(r,f) = o(T(r,f))$ outside a possible exceptional set of finite linear measure. We also note that several inequalities below, which are mostly related to Nevanlinna theory, are valid for values of r large enough, although this is not mentioned explicitly.

1.2. Valiron-Mohon'ko theorem

The following important result due to G. Valiron [22] and A. Mohon'ko [18] will be used in several occasions later on. For the proof, see [17], Theorem 2.2.5 and Corollary 2.2.7.

Theorem 1.2.1. ***(Valiron-Mohon'ko)*** *Let f be a meromorphic function. Then for all irreducible rational functions in f,*

$$R(z, f(z)) = \frac{\sum_{i=0}^{p} a_i(z) f(z)^i}{\sum_{j=0}^{q} b_j(z) f(z)^j},$$

with meromorphic coefficients $a_i(z), b_j(z)$, the characteristic function of $R(z, f(z))$ satisfies

$$T(r, R(z, f(z))) = dT(r,f) + O(\psi(r)) + S(r,f),$$

where $d = \max\{p,q\}$ and

$$\psi(r) = \max_{i,j} \{T(r,a_i), T(r,b_j)\}.$$

1.3. Some results for composite functions

The following two lemmas are due to R. Goldstein. They appear to be very useful while considering the growth of composite functions.

Lemma 1.3.1 ([11], p. 362–363). *Let f be a transcendental meromorphic function and $p(z) = a_k z^k + a_{k-1} z^{k-1} + \cdots + a_1 z + a_0$, $a_k \neq 0$, $k \geq 1$, be a polynomial of degree k. Given $0 < \delta < |a_k|$, let*

$$\begin{cases} \lambda = |a_k| + \delta \\ \mu = |a_k| - \delta \end{cases}.$$

Then, given $\varepsilon > 0$, for any $a \in \mathbb{C} \cup \infty$ and for r large enough, we have

$$kn(\mu r^k, a, f) \le n(r, a, f \circ p) \le kn(\lambda r^k, a, f), \tag{1.1}$$
$$N(\mu r^k, a, f) + O(\log r) \le N(r, a, f \circ p) \le N(\lambda r^k, a, f) + O(\log r), \tag{1.2}$$
$$(1-\varepsilon)T(\mu r^k, f) \le T(r, f \circ p) \le (1+\varepsilon)T(\lambda r^k, f). \tag{1.3}$$

Lemma 1.3.2 ([10], Lemma 3). *Let $\psi : [r_0, \infty) \to (0, \infty)$ be positive and bounded in every finite interval. Suppose that*

$$\psi(\mu r^m) \le A\psi(r) + B, \quad (r \ge r_0), \tag{1.4}$$

where $\mu > 0$, $m > 1$, $A > 1$ and B are real constants. Then

$$\psi(r) = O((\log r)^\alpha)$$

where

$$\alpha = \frac{\log A}{\log m}.$$

If $A = 1$ and $B > 0$, then for any $\varepsilon > 0$,

$$\psi(r) = O((\log r)^\varepsilon).$$

Suppose further, with the same notations as above, that

$$\psi(\mu r^m) \ge A\psi(r), \quad (r \ge r_0). \tag{1.5}$$

Then, for all sufficiently large values of r, there exists a constant $K > 0$ such that

$$\psi(r) \ge K(\log r)^\alpha$$

where

$$\alpha = \frac{\log A}{\log m}.$$

Remark. In the case when $A = 1$ and $B = 0$ in (1.4), we have $\psi(r) = O(1)$.

2. The equation $f(az + b) = R(z, f)$

A major part of this thesis is devoted to functional equations of the type $f(p(z)) = R(z, f)$ in the case when the polynomial $p(z)$ is of degree at least 2. There exist many recent studies about this type of equation in the case when $\deg p = 1$, see [12], [16], where $p(z) = cz$ and [13], where $p(z) = z + c$, $c \in \mathbb{C}$. In this section we give some results concerning the growth of the meromorphic solutions of the general first order case

$$f(az + b) = R(z, f),$$

where $a, b \in \mathbb{C}$ and $R(z, f)$ is an irreducible rational function in f with meromorphic coefficients which are small with respect to f.

2.1. The order of growth

We first need the following lemma:

Lemma 2.1.1 ([12], Lemma 3.1). *Let $\Phi : (1, \infty) \to (0, \infty)$ be a monotone increasing function, and let f be a nonconstant meromorphic function. If for some real constant $\alpha \in (0,1)$, there exist real constants $K_1 > 0$ and $K_2 \geq 1$ such that*

$$T(r,f) \leq K_1 \Phi(\alpha r) + K_2 T(\alpha r, f) + S(\alpha r, f), \tag{2.1}$$

then

$$\rho(f) \leq \frac{\log K_2}{-\log \alpha} + \limsup_{r\to\infty} \frac{\log \Phi(r)}{\log r}. \tag{2.2}$$

We also need two observations, due to Ishizaki [5]:

$$T(r, f(cz)) = T(|c|r, f) + O(1) \tag{2.3}$$

and to Ablowitz et al [1]:

$$T(r, f(z+c)) \leq (1+\varepsilon) T(r + |c|, f) + M, \tag{2.4}$$

where $M = M(\varepsilon)$ is a constant. Given $\varepsilon > 0$, (2.3) and (2.4) are true for any meromorphic function f and any constant $c \in \mathbb{C}$, provided r is large enough. Now suppose that f is a meromorphic solution of the equation

$$f(az+b) = R(z,f) = \frac{\sum_{i=0}^{m} a_i(z) f(z)^i}{\sum_{j=0}^{n} b_j(z) f(z)^j}, \tag{2.5}$$

where each a_i, b_j is a small meromorphic function with respect to f. Set $d = \max\{m, n\}$. Then by applying the Valiron-Mohon'ko theorem, (2.3) and (2.4), we obtain

$$\begin{aligned} dT(r,f) + S(r,f) = T(r,R) &= T\left(r, f\left(a\left(z + \frac{b}{a}\right)\right)\right) \\ &\leq (1+\varepsilon) T\left(|a|r + \left|\frac{b}{a}\right|, f\right) + S(r,f). \end{aligned} \tag{2.6}$$

It is immediate from (2.6) that if $d \geq 2$ and $|a| < 1$, then the equation (2.5) does not have a meromorphic solution. It is also proved ([13], Proposition 8) that if $a = 1$, if the coefficients a_i and b_j in (2.5) are rational and if the equation (2.5) admits a transcendental meromorphic solution of finite order, then $d = 1$.

We proceed with an upper bound estimate for the order of growth of a transcendental meromorphic solution f of (2.5) when $|a| > 1$.

Theorem 2.1.2. *Suppose that f is a transcendental meromorphic solution of (2.5), $a, b \in \mathbb{C}$, $|a| > 1$, and the coefficients a_i and b_j are of growth $S(r, f)$. Then*

$$\rho(f) \leq \frac{\log d}{\log |a|}, \tag{2.7}$$

where $d = \max\{m, n\}$.

Proof. By applying (1.3) and the Valiron-Mohon'ko theorem to equation (2.5), we obtain

$$(1-\varepsilon)T(\mu r, f) \leq T(r, f(az+b)) = dT(r, f) + S(r, f),$$

that is

$$T(\mu r, f) \leq \frac{d}{(1-\varepsilon)} T(r, f) + S(r, f),$$

where $\mu = |a| - \delta > 1$, $\delta > 0$. Denoting $\alpha := \frac{1}{\mu}$, we have $0 < \alpha < 1$, and we deduce that

$$T(r, f) \leq \frac{d}{(1-\varepsilon)} T(\alpha r, f) + S(\alpha r, f).$$

Applying Lemma 2.1.1 now yields

$$\rho(f) \leq \frac{\log \frac{d}{1-\varepsilon}}{-\log \alpha}.$$

The proof is now completed by letting $\varepsilon \to 0$ and $\delta \to 0$. □

An example of such solutions is given by

Example 2.1.3. Let $c \in \mathbb{C}$ be a constant such that $c \neq \frac{\pi}{2} m$, where $m \in \mathbb{Z}$. Since

$$\tan(2z + c) = \frac{2\tan z - C(1 - \tan^2 z)}{1 - \tan^2 z + C(2\tan z)},$$

where $C := -\tan c \neq 0, \infty$, we see that $f(z) = \tan z$ solves

$$f(2z + c) = \frac{2f(z) - C(1 - f(z)^2)}{1 - f(z)^2 + 2Cf(z)}.$$

For this example, the inequality (2.7) becomes an equality. Hence, (2.7) is sharp.

In order to obtain a lower bound estimate for the lower order $\mu(f)$ we first need the following lemma, see, for example, [2], Lemma 6.

Lemma 2.1.4. *Suppose that a meromorphic function f has finite lower order λ. Then for every constant $c > 1$ and a given ε there exists a sequence $r_n = r_n(c, \varepsilon) \to \infty$ such that*

$$T(cr_n, f) \leq c^{\lambda+\varepsilon} T(r_n, f).$$

Theorem 2.1.5. *Suppose that f is a transcendental meromorphic solution of (2.5), $a, b \in \mathbb{C}$, $|a| > 1$, and the coefficients a_i and b_j are small with respect to f. Then for the lower order $\mu(f)$, we have*

$$\frac{\log d}{\log |a|} \leq \mu(f) < \infty,$$

where $d = \max\{m, n\}$.

Proof. By (2.6), for given $\varepsilon_1 > 0$,

$$dT(r, f) \leq (1 + \varepsilon_1) T\left(|a|r + \left|\frac{b}{a}\right|, f\right) + S(r, f).$$

There is a constant $c = |a| + \varepsilon_2$, $\varepsilon_2 > 0$, such that

$$T\left(|a|r + \left|\frac{b}{a}\right|, f\right) \leq T(cr, f),$$

and therefore, when r is large enough,

$$dT(r, f) \leq (1 + \varepsilon_3) T(cr, f),$$

outside a possible exceptional set of finite linear measure. By the Bank lemma, see, [17], Lemma 1.1.1, for given $\sigma > 1$,

$$dT(r, f) \leq (1 + \varepsilon_3) T(\sigma cr, f)$$

for all r large enough. Set $\lambda = \mu(f)$. From Theorem 2.1.2, we have $\lambda < \infty$. Then from Lemma 2.1.4, for any given $\varepsilon_4 > 0$, there exists a sequence $r_n = r_n(c, \varepsilon_4) \to \infty$ such that for $r_n > r_0$

$$T(cr_n, f) \leq c^{\lambda + \varepsilon_4} T(r_n, f).$$

Hence

$$dT(r_n, f) \leq (1 + \varepsilon_3)(\sigma c)^{\lambda + \varepsilon_4} T(r_n, f),$$

which implies

$$d \leq (1 + \varepsilon)(\sigma c)^{\lambda + \varepsilon}, \quad \varepsilon = \max\{\varepsilon_1, \varepsilon_2, \varepsilon_3, \varepsilon_4\}, \tag{2.8}$$

by taking lower limit as $n \to \infty$. Since the lower bound d is a constant and (2.8) holds for any $\varepsilon > 0$, we may let $\varepsilon \to 0$ and $\sigma \to 1$ to obtain

$$d \leq |a|^{\lambda},$$

from which we deduce

$$\lambda \geq \frac{\log d}{\log |a|}.$$

□

The next corollary immediately follows from the two preceding theorems:

Corollary 2.1.6. *Transcendental meromorphic solutions of* (2.5), *where* $|a| > 1$, *are of regular growth.*

Theorem 2.1.7. *Suppose that* f *is a meromorphic solution of an equation of the form* (2.5), *where* $|a| > 1$. *If* $\rho := \max_{i,j}\{\rho(a_i), \rho(b_j)\}$, *then*

$$\rho(f) \leq \max\left\{\rho, \frac{\log d}{\log|a|}\right\}. \tag{2.9}$$

Proof. We may assume that $\rho < \infty$. Furthermore, by assuming $\rho < \rho(f)$, we prove that

$$\rho(f) \leq \frac{\log d}{\log|a|}. \tag{2.10}$$

Then (2.9) immediately follows.

To this end, let β and γ be fixed constants satisfying $\rho < \beta < \gamma < \rho(f)$, and set $\alpha := \frac{1}{\mu}$, where $\mu = |a| - \delta > 1$. Denote $\phi(r) := \max_{i,j}\{T(r, a_i), T(r, b_j)\}$. By applying (1.3) and the Valiron-Mohon'ko theorem to equation (2.5) we deduce similarly as earlier in the proof of Theorem 2.1.2 that

$$T(r, f) \leq \frac{d}{(1-\varepsilon)} T(\alpha r, f) + \phi(\alpha r). \tag{2.11}$$

where, for some constant $K > 0$,

$$|\phi(r)| \leq K r^{\beta}$$

for all r large enough. If

$$I := \{r \geq 0 \mid T(r, f) > r^{\gamma}\},$$

then clearly,

$$\rho(f) = \limsup_{r\to\infty, r\in I} \frac{\log T(r, f)}{\log r}.$$

Now fix an $R > 0$. For each $r \geq R$, there exists an integer $m = m(r) \geq 1$ such that

$$R\mu^{m-1} \leq r < R\mu^{m}. \tag{2.12}$$

From (2.12),

$$\lim_{r\to\infty} \frac{m}{\log r} = \frac{1}{\log \mu}. \tag{2.13}$$

We next use induction on (2.11) to obtain

$$T(r, f) \leq \left(\frac{d}{1-\varepsilon}\right)^{m} T(\alpha^{m} r, f) + \nu(r), \tag{2.14}$$

where

$$\nu(r) = \sum_{k=0}^{m-1} \left(\frac{d}{1-\varepsilon}\right)^k \phi(\alpha^{k+1} r).$$

Since $|\phi(r)| \leq K r^\beta$ for all r large enough,

$$|\nu(r)| \leq K \sum_{k=0}^{m-1} \left(\frac{d}{1-\varepsilon}\right)^k (\alpha^{k+1} r)^\beta \leq K r^\beta \sum_{k=0}^{m-1} \left(\frac{d}{1-\varepsilon}\alpha^\beta\right)^k.$$

Case (i): $\frac{d}{1-\varepsilon}\alpha^\beta > 1$. This gives

$$\beta < \frac{\log \frac{d}{1-\varepsilon}}{\log \mu}.$$

Since β can be chosen arbitrarily close to $\rho(f)$, this yields (2.10), letting $\varepsilon \to 0$ and $\delta \to 0$.

Case (ii): $\frac{d}{1-\varepsilon}\alpha^\beta \leq 1$. Then $|\nu(r)| \leq K m r^\beta$, and so from (2.13) and $\beta < \gamma$, we obtain

$$\nu(r) = o\left(r^\gamma\right).$$

Then from (2.14) and the definition of I, we have

$$(1 + o(1)) T(r, f) \leq \left(\frac{d}{1-\varepsilon}\right)^m T(\alpha^m r, f), \quad \text{as } r \to \infty,\ r \in I.$$

Hence from (2.12) and (2.13),

$$\begin{aligned} \rho(f) &= \limsup_{r\to\infty, r\in I} \frac{\log T(r, f)}{\log r} \leq \limsup_{r\to\infty, r\in I} \frac{\log((\frac{d}{1-\varepsilon})^m T(\alpha^m r, f))}{\log r} \\ &\leq \limsup_{r\to\infty, r\in I} \frac{m \log \frac{d}{1-\varepsilon} + \log T(R, f)}{\log r} = \frac{\log \frac{d}{1-\varepsilon}}{\log \mu}, \end{aligned}$$

proving (2.10) in this final case. □

2.2. A special case

In this section we briefly consider a special case of the equation (2.5), namely, the equation

$$f(az + b) = A(z) + \sum_{k=1}^{m} a_k f(z)^k, \tag{2.15}$$

where $A(z)$ is an entire function, $a \in \mathbb{C}$, $|a| > 1$, $b \in \mathbb{C}$, $b \neq 0$, $m \in \mathbb{N}$, $a_k \in \mathbb{C}$ and $a_m \neq 0$.

Theorem 2.2.1. *Every meromorphic solution of* (2.15) *is entire.*

Proof. Let f be a meromorphic solution of (2.15) and suppose that z_0 is a pole of f. If $z_0 = 0$, then we can always find a non-zero pole, $z_1 = -\frac{b}{a} \neq 0$. So, we can assume, without a loss of generality, that $z_0 \neq 0$. By (2.15) f must have a pole at $\frac{1}{a}z_0 - \frac{b}{a}$. Continuing this process, we obtain that f has a pole at $\frac{1}{a^2}z_0 - \frac{b}{a^2} - \frac{b}{a}$ and inductively, at $\frac{1}{a^m}z_0 - b\sum_{k=1}^{m}\frac{1}{a^k}$ for any $m \in \mathbb{N}$. Since $\frac{1}{a^m}z_0 \to 0$ and $-b\sum_{k=1}^{m}\frac{1}{a^k}$ goes to a nonzero constant, that constant is the limit point of poles of f, a contradiction. Hence f is entire. □

3. The equation $f(p(z)) = R(z, f)$

We now proceed to consider the general equation

$$f(p(z)) = R(z, f(z)) = \frac{\sum_{i=0}^{m} a_i(z) f(z)^i}{\sum_{j=0}^{n} b_j(z) f(z)^j}, \tag{3.1}$$

where $p(z) = c_k z^k + c_{k-1}z^{k-1} + \cdots + c_1 z + c_0$ is a polynomial of degree $k \geq 2$, $R(z, f)$ is an irreducible rational function in f and $d = \max\{m, n\}$. In this chapter we assume that f is a transcendental meromorphic function and the meromorphic coefficients a_i and b_j are of growth $S(r, f)$ unless otherwise specified. Furthermore, we assume that a_m and b_n do not vanish identically.

The first step in generalizing the equation (3.1) would be replacing the polynomial p with an entire function g. We first need the following

Lemma 3.1. ([3], p. 35–36) *Let f be a non-constant meromorphic function and let g be a transcendental entire function. Then there exists an increasing sequence, $r_n \to \infty$, such that*

$$T(r, f \circ g) \geq T\left(\left(M\left(\frac{r}{4}, g\right)\right)^{\frac{1}{30}}, f\right) \tag{3.2}$$

holds for $r = r_n$.

Theorem 3.2. *Let f be a non-constant meromorphic solution of the equation*

$$f(g(z)) = R(z, f(z)) = \frac{\sum_{i=0}^{m} a_i(z) f(z)^i}{\sum_{j=0}^{n} b_j(z) f(z)^j}, \tag{3.3}$$

where g is an entire function, $R(z, f)$ is an irreducible rational function in f and the coefficients a_i and b_j are small meromorphic functions with respect to f. Then g is a polynomial.

Proof. Assume that g is transcendental. By (3.3) and the Valiron-Mohon'ko theorem

$$T(r, f \circ g) = T(r, R) \leq d(1+\varepsilon)T(r, f), \quad d = \max\{m, n\},$$

outside a possible exceptional set of finite linear measure. By the Bank lemma

$$T(r, f \circ g) \leq d(1+\varepsilon)T(\alpha r, f), \quad \alpha > 1, \quad r \geq r_\alpha. \tag{3.4}$$

Since g is a transcendental entire function

$$\liminf_{r\to\infty} \frac{\log M(r,g)}{\log r} = \infty.$$

Hence for any given $K > 30$ and for r large enough

$$M(r, g) > r^K.$$

Therefore, there exists an increasing sequence $r_n \to \infty$, as in Lemma 3.1, such that

$$M\left(\frac{r_n}{4}, g\right) > \left(\frac{r_n}{4}\right)^K$$

for any n. By applying (3.2) we obtain

$$T(r_n, f \circ g) \geq T\left(\left(\frac{r_n}{4}\right)^{\frac{K}{30}}, f\right). \tag{3.5}$$

Note that $T(r, f)/\log r$ is an increasing function of r. Since

$$\left(\frac{r_n}{4}\right)^{\frac{K}{30}} > \alpha r_n,$$

for sufficiently large n, we have

$$T\left(\left(\frac{r_n}{4}\right)^{\frac{K}{30}}, f\right) > \frac{\frac{K}{30}(\log r_n - \log 4)}{\log r_n + \log \alpha} T(\alpha r_n, f) > \frac{K}{40} T(\alpha r_n, f), \tag{3.6}$$

as $n \to \infty$. By (3.4), (3.5) and (3.6) we obtain

$$d(1+\varepsilon)T(\alpha r_n, f) \geq T\left(\left(\frac{r_n}{4}\right)^{\frac{K}{30}}, f\right) > \frac{K}{40} T(\alpha r_n, f),$$

as $n \to \infty$. Since K can be arbitrarily large, this is a contradiction. So, g must be a polynomial. □

3.1. The case $d = 1$

When we consider the existence of transcendental meromorphic solutions in Sections 5 and 6, a special case of the next theorem appears to be useful.

Theorem 3.1.1. *The functional equation*

$$f(p(z)) = \frac{a_0(z) + a_1(z)f(z)}{b_0(z) + b_1(z)f(z)}, \tag{3.7}$$

where the coefficients a_i and b_j are of growth $S(r,f)$ such that $a_0(z)b_1(z) - a_1(z)b_0(z) \not\equiv 0$ and $p(z) = c_k z^k + c_{k-1}z^{k-1} + \cdots + c_1 z + c_0$ is a polynomial of $\deg p = k \geq 2$, does not have meromorphic solutions.

Proof. Assume first that there is a transcendental meromorphic solution f. For a sufficiently small $\varepsilon > 0$, we have by the Valiron-Mohon'ko theorem and (1.3) that outside a possible exceptional set of finite linear measure

$$(1-\varepsilon)T(\mu r^k, f) \leq T(r, f \circ p) \leq (1+\varepsilon)T(r,f),$$

where $\mu = |c_k|(1-\varepsilon)$. Hence, for $\alpha > 1$ and for r large enough

$$(1-\varepsilon)T(\mu r^k, f) \leq (1+\varepsilon)T(\alpha r, f).$$

Set $\alpha r = t$. Then

$$T\left(\frac{\mu}{\alpha^k}t^k, f\right) \leq \frac{(1+\varepsilon)}{(1-\varepsilon)}T(t,f).$$

By Lemma 1.3.2 we get

$$T(t,f) = O(\log t)^\alpha,$$

where

$$\alpha = \frac{\log\frac{(1+\varepsilon)}{(1-\varepsilon)}}{\log k} < 1$$

a contradiction.

Assume next that f is rational. Then the coefficients must be constants. Suppose that $\deg f = m$. Now

$$\deg(f \circ p) = km = \deg\left(\frac{a_0(z) + a_1(z)f(z)}{b_0(z) + b_1(z)f(z)}\right) \leq \deg f = m.$$

So, $k \leq 1$, a contradiction. □

From Theorem 3.1.1, equation (3.7) does not have a meromorphic solution when the coefficients are small meromorphic functions with respect to f. The next example shows that at least if the coefficients a_i and b_j are rational, there may exist a rational solution of (3.7).

Example 3.1.2. The function $f(z) = \frac{1}{1-z}$ is a solution of the equation

$$f(z^2) = \frac{\frac{1}{2} + \frac{1}{2}(1-z)f(z)}{(1-z)f(z) - z^2}.$$

3.2. Growth of meromorphic solutions

By considering the characteristic function of a transcendental meromorphic solution of the general equation (3.1) we obtain that the solution f is of regular growth.

Theorem 3.2.1. *Suppose that f is a transcendental meromorphic solution of*

$$f(p(z)) = R(z, f(z)) = \frac{\sum_{i=0}^{m} a_i(z) f(z)^i}{\sum_{j=0}^{n} b_j(z) f(z)^j},$$

where $p(z) = c_k z^k + c_{k-1} z^{k-1} + \cdots + c_1 z + c_0$ is a polynomial of degree $k \geq 2$, $R(z, f)$ is an irreducible rational function in f such that $d = \max\{m, n\} \geq k$ and $T(r, a_i)$, $T(r, b_j) = S(r, f)$. Then

$$T(r, f) = O((\log r)^{\alpha+\varepsilon})$$

where

$$\alpha = \frac{\log d}{\log k}, \qquad 0 < \varepsilon < 1.$$

Proof. For $\varepsilon > 0$, we have by the Valiron-Mohon'ko theorem and (1.3) that outside a possible exceptional set of finite linear measure

$$(1-\varepsilon) T(\mu r^k, f) \leq T(r, f \circ p) \leq T(r, R(z, f)) \leq d(1+\varepsilon) T(r, f),$$

where $\mu = |c_k|(1-\varepsilon)$. Therefore, for $\alpha > 1$ and for all r large enough

$$T(\mu r^k, f) \leq d \frac{(1+\varepsilon)}{(1-\varepsilon)} T(\alpha r, f).$$

Hence, similarly as in the proof of Theorem 3.1.1

$$T\left(\frac{\mu}{\alpha^k} t^k, f\right) \leq d \frac{(1+\varepsilon)}{(1-\varepsilon)} T(t, f),$$

and by Lemma 1.3.2 we have

$$T(t, f) = O((\log t)^{\alpha_1}),$$

where

$$\alpha_1 = \frac{\log\left(d \frac{(1+\varepsilon)}{(1-\varepsilon)}\right)}{\log k} = \frac{\log d + \log \frac{(1+\varepsilon)}{(1-\varepsilon)}}{\log k} = \frac{\log d}{\log k} + \varepsilon_1.$$

The assertion follows. □

Theorem 3.2.1 immediately gives the next result.

Corollary 3.2.2. *Suppose that f is a transcendental meromorphic solution of*

$$f\left(p(z)\right) = R\left(z, f(z)\right) = \frac{\sum_{i=0}^{m} a_i(z) f(z)^i}{\sum_{j=0}^{n} b_j(z) f(z)^j},$$

where $p(z) = c_k z^k + c_{k-1} z^{k-1} + \cdots + c_1 z + c_0$ is a polynomial of degree $k \geq 2$, $R\left(z, f(z)\right)$ is an irreducible rational function in f such that $d = \max\{m, n\} \geq k$ and $T(r, a_i)$, $T(r, b_j) = S(r, f)$. Then

$$\rho(f) = \mu(f) = 0.$$

Theorem 3.2.3. *Suppose that f is a transcendental meromorphic solution of*

$$f\left(p(z)\right) = R\left(z, f(z)\right) = \frac{\sum_{i=0}^{m} a_i(z) f(z)^i}{\sum_{j=0}^{n} b_j(z) f(z)^j},$$

where $p(z) = c_k z^k + c_{k-1} z^{k-1} + \cdots + c_1 z + c_0$ is a polynomial of degree $k \geq 2$, $R\left(z, f(z)\right)$ is an irreducible rational function in f such that $d = \max\{m, n\} \geq k$ and $T(r, a_i)$, $T(r, b_j) = S(r, f)$. Then for a positive constant K

$$T(r, f) \geq K(\log r)^{\alpha - \varepsilon},$$

where

$$\alpha = \frac{\log d}{\log k}, \qquad \varepsilon > 0.$$

Proof. For $\varepsilon > 0$ we have by the Valiron-Mohon'ko theorem and (1.3) that outside a possible exceptional set of finite linear measure

$$(1 + \varepsilon) T(\lambda r^k, f) \geq T(r, f \circ p) = T(r, R(z, f)) \geq d(1 - \varepsilon) T(r, f),$$

where $\lambda = |c_k|(1 + \varepsilon)$. Hence for $\alpha > 1$ and for all r large enough

$$(1 + \varepsilon) T(\alpha \lambda r^k, f) \geq d(1 - \varepsilon) T(r, f).$$

Writing this

$$T(\lambda r^k, f) \geq d \frac{(1 - \varepsilon)}{(1 + \varepsilon)} T(r, f)$$

and using Lemma 1.3.2 we get

$$T(r, f) \geq K(\log r)^{\alpha_1},$$

where

$$\alpha_1 = \frac{\log\left(d \frac{(1-\varepsilon)}{(1+\varepsilon)}\right)}{\log k} = \frac{\log d + \log \frac{(1-\varepsilon)}{(1+\varepsilon)}}{\log k} = \frac{\log d}{\log k} - \varepsilon_1.$$

The assertion follows. □

Combining Theorems 3.2.1 and 3.2.3 we have

Corollary 3.2.4. *Suppose that f is a transcendental meromorphic solution of*

$$f(p(z)) = R(z, f(z)) = \frac{\sum_{i=0}^{m} a_i(z) f(z)^i}{\sum_{j=0}^{n} b_j(z) f(z)^j},$$

where $p(z) = c_k z^k + c_{k-1} z^{k-1} + \cdots + c_1 z + c_0$ is a polynomial of degree $k \geq 2$, $R(z, f(z))$ is an irreducible rational function in f such that $d = \max\{m, n\} \geq k$ and $T(r, a_i), T(r, b_j) = S(r, f)$. Then

$$K_1 (\log r)^{\alpha - \varepsilon} \leq T(r, f) \leq K_2 (\log r)^{\alpha + \varepsilon},$$

where

$$\alpha = \frac{\log d}{\log k}, \qquad 0 < \varepsilon < 1,$$

and K_i, $i = 1, 2$, are positive constants.

3.3. Meromorphic solutions with few zeros and poles

The following result shows that non-rational solutions having few zeros and poles appear in special situations only.

Theorem 3.3.1. *Suppose that f is a transcendental meromorphic solution of*

$$f(p(z)) = R(z, f(z)) = \frac{\sum_{i=0}^{m} a_i(z) f(z)^i}{\sum_{j=0}^{n} b_j(z) f(z)^j}, \tag{3.8}$$

with meromorphic coefficients a_i, b_j of growth $S(r, f)$ such that $a_m(z) b_n(z) \not\equiv 0$. The polynomial $p(z) = c_k z^k + c_{k-1} z^{k-1} + \ldots + c_1 z + c_0$ is of degree $k \geq 2$. If

$$\overline{N}(r, f \circ p) + \overline{N}\left(r, \frac{1}{f \circ p}\right) = S(r, f), \tag{3.9}$$

then (3.8) is of the form

$$f(p(z)) = c(z) f(z)^s, \tag{3.10}$$

where $c(z)$ is meromorphic, $T(r, c) = S(r, f)$ and $s \in \mathbb{Z}$, $s \neq 0$.

Proof. Let f be transcendental and let z_0 be a zero of f. Then there exist at most t points, $\zeta_1, \ldots, \zeta_t$, $t \leq k$, such that $p(\zeta_j) = z_0$ and $f(p(\zeta_j)) = 0$. Hence

$$t\overline{N}\left(r, \frac{1}{f}\right) \leq \overline{N}\left(r, \frac{1}{f \circ p}\right).$$

Therefore, by using (3.9) and the lemma of the logarithmic derivative we get

$$T\left(r, \frac{f'}{f}\right) = \overline{N}(r, f) + \overline{N}\left(r, \frac{1}{f}\right) + S(r, f) = S(r, f)$$

and since $S(r, f\circ p) = S(r,R) = S(r,f)$

$$T\left(r,\frac{(f\circ p)'}{f\circ p}\right) = N\left(r,\frac{(f\circ p)'}{f\circ p}\right) + m\left(r,\frac{(f\circ p)'}{f\circ p}\right)$$
$$\le \overline{N}(r,f\circ p) + \overline{N}\left(r,\frac{1}{f\circ p}\right) + S(r,f) = S(r,f).$$

We may write (3.8) in the form

$$\frac{b_n(z)}{a_m(z)} f(p(z)) = \frac{P(z,f)}{Q(z,f)} = u(z,f). \tag{3.11}$$

Clearly P and Q are monic polynomials in f with coefficients of growth $S(r,f)$. Denote $F := \frac{f'}{f}$, $U := \frac{u'}{u}$ and notice that $T(r,U) = S(r,f)$ by (3.11). Since

$$\frac{P'Q - PQ'}{Q^2} = u' = Uu = \frac{UP}{Q},$$

we get

$$P'Q - PQ' = UPQ. \tag{3.12}$$

Writing $f' = Ff$ in (3.12), and then regarding (3.12) as an algebraic equation in f with coefficients of growth $S(r,f)$ and comparing the leading coefficients, we obtain by [15]

$$(m-n)F = U.$$

Therefore,

$$u(z,f) = \alpha f(z)^{m-n}$$

for some $\alpha \in \mathbb{C}$, and so

$$f(p(z)) = \alpha \frac{a_m(z)}{b_n(z)} f(z)^{m-n}, \tag{3.13}$$

proving the assertion. □

We will study the solutions of the equation (3.10) in Section 5 in more detail.

4. The equation $f(p(z)) = a(z)R(z,f)$

Before considering more special cases we study another equation of rather general type, namely,

$$f(p(z)) = a(z)R(z,f),$$

where the functions f and a are meromorphic and $R(z,f)$ is an irreducible rational function in f such that the coefficients a_i and b_j are small meromorphic functions with respect to f.

4.1. Growth relations

Theorem 4.1.1. *Suppose that f is a transcendental meromorphic solution of the equation*

$$f(p(z)) = a(z)R(z,f) = a(z)\frac{\sum_{i=0}^{m} a_i(z)f(z)^i}{\sum_{j=0}^{n} b_j(z)f(z)^j}, \tag{4.1}$$

where a is meromorphic, $R(z,f)$ is an irreducible rational function in f such that the coefficients a_i and b_j are small meromorphic functions with respect to f and p is a polynomial, $\deg p = k \geq 2$. Then, provided that

$$\limsup_{r\to\infty} \frac{T(r,f)}{(\log r)^K} = \infty$$

for all $K > 1$, we have

$$\limsup_{r\to\infty} \frac{T(r,a)}{T(r,f)} = \infty.$$

Proof. Using the properties of the characteristic function, we obtain from (4.1) that

$$T(r, f\circ p) \leq T(r,a) + T(r,R). \tag{4.2}$$

Also, rewriting (4.1) as

$$a(z) = f(p(z))\frac{1}{R(z,f)},$$

we obtain

$$T(r, f\circ p) \geq T(r,a) - T(r,R) + O(1). \tag{4.3}$$

From the Valiron-Mohon'ko theorem we know that $T(r,R) = dT(r,f) + S(r,f)$, where $d = \max\{m,n\}$. Hence, by (4.2) and (4.3), we get outside a possible exceptional set of finite linear measure, that

$$\left| \frac{T(r,f\circ p)}{T(r,f)} - \frac{T(r,a)}{T(r,f)} \right| \leq C, \tag{4.4}$$

where C is a positive constant. Therefore, it is sufficient to prove that

$$\limsup_{r\to\infty} \frac{T(r,f\circ p)}{T(r,f)} = \infty \tag{4.5}$$

provided

$$\limsup_{r\to\infty} \frac{T(r,f)}{(\log r)^K} = \infty$$

for all $K > 1$. Let us assume on the contrary that

$$\limsup_{r\to\infty} \frac{T(r,f\circ p)}{T(r,f)} < \infty \tag{4.6}$$

and

$$\limsup_{r\to\infty}\frac{T(r,f)}{(\log r)^K}=\infty$$

for all $K>1$. By (4.6) and (1.3), for $\alpha>1$ and for r large enough

$$(1-\varepsilon)T(\mu r^k,f)\le T(r,f\circ p)\le MT(\alpha r,f),$$

where M is a constant. Then,

$$T\left(\frac{\mu}{\alpha^k}r^k,f\right)\le\frac{M}{(1-\varepsilon)}T(r,f)$$

and by Lemma 1.3.2

$$T(r,f)=O(\log r)^{\frac{\log M}{\log k}}.$$

Hence

$$\limsup_{r\to\infty}\frac{T(r,f)}{(\log r)^{\frac{\log M}{\log k}}}<\infty,$$

a contradiction. □

5. The equation $f(p(z))=a(z)f(z)^m$

This section is devoted to studying the equation

$$f(p(z))=a(z)f(z)^m, \tag{5.1}$$

where f and a are meromorphic, p is a polynomial with $\deg p=k\ge 2$ and $m\in\mathbb{Z}$. We prove that there do not exist any transcendental meromorphic solutions if a is a rational function. Further, we prove a result concerning the solutions of the equation (5.1) when $a=Se^Q$, where S is rational and Q is a polynomial.

5.1. The existence of meromorphic solutions when $a(z)$ is rational

Using logarithmic derivatives we reduce the equation (5.1) to the case $f(p(z))=A(z)f(z)+B(z)$, where A and B are small meromorphic functions with respect to f. To this end we need the following

Lemma 5.1.1. *Suppose that f is a meromorphic solution of the equation*

$$f(p(z))=A(z)f(z)+B(z), \tag{5.2}$$

where p is a polynomial of degree $k\ge 2$. If A and B are rational functions, then f is a rational function.

Proof. See Theorem 3.1.1. □

Theorem 5.1.2. *Suppose that f is a meromorphic solution of the equation*

$$f(p(z)) = a(z)f(z)^m, \tag{5.3}$$

where a is a rational function, m is an integer, $p(z) = c_k z^k + c_{k-1}z^{k-1} + \cdots + c_0$ and $\deg p = k \geq 2$. Then f must be rational.

Proof. From (5.3) we get

$$g(p(z)) = \frac{m}{p'(z)}g(z) + \frac{1}{p'(z)}\frac{a'(z)}{a(z)}, \tag{5.4}$$

where $g = \frac{f'}{f}$. By Lemma 5.1.1, (5.4) does not have transcendental meromorphic solutions, which means that g must be rational. Since we assumed that f is meromorphic, the poles of the rational function g are exactly at the zeros and the poles of f. So, by the Weierstrass factorization,

$$f(z) = H(z)e^{P(z)},$$

where H is a rational and P is an entire function. Then, from

$$g(z) = \frac{f'(z)}{f(z)} = \frac{H'(z)}{H(z)} + P'(z)$$

we see that P' is a polynomial. We can write g in a form

$$g(z) = \sum_{j=1}^{n} \frac{\alpha_j}{z - z_j} + Q(z), \tag{5.5}$$

where $\alpha_j \in \mathbb{Z}$, $j = 1, \ldots, n$ and $Q := P'$ is a polynomial. We write $a = \frac{P_1}{Q_1}$, where P_1 and Q_1 are polynomials. Now

$$\frac{a'(z)}{a(z)} = \frac{P_1'}{P_1} - \frac{Q_1'}{Q_1} = \sum_{k=1}^{s} \frac{\gamma_k}{z - \beta_k}, \tag{5.6}$$

where $s \leq \deg P_1 + \deg Q_1$. Substituting (5.5) and (5.6) into (5.4) we have

$$\sum_{j=1}^{n} \frac{\alpha_j}{p(z) - z_j} + Q(p(z)) = \frac{m}{p'(z)} \sum_{j=1}^{n} \frac{\alpha_j}{z - z_j} + \frac{m}{p'(z)}Q(z) + \frac{1}{p'(z)} \sum_{k=1}^{s} \frac{\gamma_k}{z - \beta_k}.$$

Letting $z \to \infty$ we deduce that

$$\lim_{z \to \infty} (p'(z)Q(p(z)) - mQ(z)) = 0. \tag{5.7}$$

By considering the degree of Q in the equation (5.7) we see that $Q = 0$. Since $P' = Q$, $\deg P = 0$ proving the assertion. □

Example 5.1.3. We give three examples of the equation (5.3) such that a is a constant, a polynomial and a rational function. First we have an example of the case $k = m$ by,

$$f(z^2 - 2z + 2) = f(z)^2, \qquad f(z) = \frac{1}{z-1}.$$

Next, the case $k < m$ is given by

$$f(z^2) = zf(z)^3, \qquad f(z) = \frac{1}{z}.$$

Finally, we have the case $k > m$ by

$$f(z^3) = \frac{z-1}{z^2+z+1} f(z)^2, \qquad f(z) = \frac{1}{z-1}.$$

5.2. Meromorphic solutions when $a(z) = S(z)e^{Q(z)}$

We consider the solutions of the equation

$$f(p(z)) = a(z)f(z)^m, \quad \text{where } m \text{ is a non-zero integer,}$$

in the case when $a(z) = S(z)e^{Q(z)}$, where S is a rational function and Q is a polynomial. The two cases $m > 0$ and $m < 0$ differ slightly and are to be considered separately.

Theorem 5.2.1. *Suppose that f is a non-trivial meromorphic solution of the equation*

$$f(p(z)) = a(z)f(z)^m, \tag{5.8}$$

where a is a non-trivial meromorphic function, p is a polynomial of degree $k \geq 2$ and m is an integer. Then

$$f(z) = H(z)e^{P(z)},$$

where H is a rational function and P is a polynomial, if and only if

$$a(z) = S(z)e^{Q(z)},$$

where S is a rational function and Q is a polynomial. Furthermore, if H is a constant then S is a constant as well.

Proof. From (5.8) we get

$$p'(z)g(p(z)) - mg(z) = \frac{a'(z)}{a(z)}, \tag{5.9}$$

where $g = \frac{f'}{f}$. Assume first that $f(z) = H(z)e^{P(z)}$, where H is a rational function and P is a polynomial. Then g is a rational function. By (5.9), $\frac{a'(z)}{a(z)}$ must also be a rational function and therefore it has finitely many zeros and poles. So, by the Weierstrass factorization a must have a form $a(z) = S(z)e^{Q(z)}$, where S is a rational function and Q is entire. Now

$$\frac{a'(z)}{a(z)} = \frac{S'(z)}{S(z)} + Q'(z)$$

and we see that Q must be a polynomial.

Suppose next that $a(z) = S(z)e^{Q(z)}$, where S is a rational function and Q is a polynomial. Then $\frac{a'(z)}{a(z)}$ is a rational function. By applying Lemma 5.1.1 to equation (5.9), we obtain that g is also a rational function. By Weierstrass factorization, $f(z) = H(z)e^{P(z)}$, where H is a rational function and P is entire. As above we see that P is a polynomial.

We see that if H is a constant, then

$$g(z) = \frac{H'(z)}{H(z)} + P'(z) = P'(z)$$

is a polynomial, and by (5.9),

$$\frac{a'(z)}{a(z)} = \frac{S'(z)}{S(z)} + Q'(z) \tag{5.10}$$

cannot have poles, that is S must be a constant. □

Remark. In the above theorem $\limsup_{r\to\infty} \frac{T(r,a)}{T(r,f)} = \infty$, which immediately follows from Theorem 4.1.1.

Remark. It is easy to find a counterexample showing that if S is a constant, it does not imply that H is also a constant. Suppose that S is a constant. Then from (5.10) we see that $\frac{a'(z)}{a(z)}$ is a polynomial. By (5.9),

$$p'(z)g(p(z)) - mg(z) \tag{5.11}$$

is also a polynomial.

Next we give an example of the non-constant H such that (5.11) is still a polynomial. If $f(z) = \frac{e^z}{z}$, then $g(z) = \frac{z-1}{z}$. Further, assume that $p(z) = z^2$ and $m = 2$. Then

$$p'(z)g(p(z)) - mg(z) = 2z\frac{z^2-1}{z^2} - 2\frac{z-1}{z} = 2z - 2.$$

We have the following

Example 5.2.2. The function $f(z) = \frac{e^z}{z}$ is a solution of the equation

$$f(z^2) = e^{z^2-2z} f(z)^2.$$

In this case, $S(z) = 1$ but $H(z) = \frac{1}{z}$.

Example 5.2.3. All functions of the form $f(z) = \frac{e^z}{z^n}$, $n \in \mathbb{N} \cup \{0\}$, are solutions of the equations

$$f(z^k) = e^{z^k - kz} f(z)^k,$$

where $k \in \mathbb{N} \setminus \{1\}$.

Theorem 5.2.4. *Suppose that $f(z) = H(z)e^{P(z)}$, where H is a rational function and P is a polynomial, is a solution of the equation*

$$f(p(z)) = \frac{a(z)}{f(z)^m}, \tag{5.12}$$

where $a(z) = S(z)e^{Q(z)}$, where S is a rational function and Q is a polynomial, p is a polynomial of degree $k \geq 2$ and m is an integer, $m > 0$. If S is a constant, then H is a constant as well.

Proof. We assume on the contrary that when S is a constant H is a non-constant rational function. We write the rational function H in the form $H = \frac{U}{V}$, where U and V are polynomials. Under the same assumptions for f and a as in Theorem 5.2.1 we obtain from (5.12) that

$$\frac{U(p(z))}{V(p(z))} e^{P(p(z))} = S e^{Q(z) - mP(z)} \frac{V(z)^m}{U(z)^m}$$

and so

$$e^{P(p(z))} = S e^{Q(z) - mP(z)} \frac{V(z)^m V(p(z))}{U(z)^m U(p(z))}. \tag{5.13}$$

In the equation (5.13) the left side is entire and it does not have any zeros. It follows that V^m and $U \circ p$, respectively U^m and $V \circ p$, share the same zeros counting multiplicity. So, we have

$$V^m = \alpha (U \circ p), \alpha \in \mathbb{C} \setminus \{0\}$$

and

$$U^m = \beta (V \circ p), \beta \in \mathbb{C} \setminus \{0\}$$

respectively. Now

$$H^m = \left(\frac{U}{V} \right)^m = \frac{\beta}{\alpha} \frac{1}{H \circ p},$$

that is

$$H(z)^m H(p(z)) = \frac{\beta}{\alpha}. \tag{5.14}$$

By (5.14),

$$\lim_{z\to\infty} H(z) = \gamma,$$

where γ is a nonzero constant. Hence, $\deg V = \deg U$ and so the number of the zeros and the poles of H is the same. We may write

$$H(z) = \gamma \frac{(z-\alpha_1)\dots(z-\alpha_s)}{(z-\beta_1)\dots(z-\beta_s)}, \quad \gamma \in \mathbb{C}\setminus\{0\}. \tag{5.15}$$

By combining (5.15) with (5.14) we have

$$\gamma^{m+1}\frac{(z-\alpha_1)^m\dots(z-\alpha_s)^m}{(z-\beta_1)^m\dots(z-\beta_s)^m}\cdot\frac{(p(z)-\alpha_1)\dots(p(z)-\alpha_s)}{(p(z)-\beta_1)\dots(p(z)-\beta_s)} = \frac{\beta}{\alpha} \neq 0, \infty. \tag{5.16}$$

Since (5.16) is a constant, there must be a connection between the poles of H and the zeros of $H \circ p$. In fact, the point β_j, which is a pole of H, is a zero, say α_k, of $H \circ p$. So, $p(\beta_j) = \alpha_k$. Assume that β_j is a pole of H of multiplicity μ. From (5.16) we see that the corresponding zero of $H \circ p$ must be of multiplicity μm. Since $p(\beta_j) = \alpha_k$, we conclude that H has a zero of multiplicity μm at α_k. Now the corresponding pole β_i is of order μm^2. This process continues until we have for some pole or zero two different multiplicities μm^l and μm^n, $l \neq n$. So, $m = 1$, a contradiction. □

5.3. On the poles of the transcendental functions f and a

These results can be found from [10], given without proofs there. For completeness, we include complete proofs below.

Lemma 5.3.1. *Suppose that f is a transcendental meromorphic solution of the equation*

$$f(p(z)) = a(z)q(f(z)), \tag{5.17}$$

where $a(z)$ is a meromorphic function, $q(z)$ is a polynomial, $\deg q = m$ and $p(z) = c_k z^k + c_{k-1}z^{k-1} + \dots + c_0$, $k \geq 2$. Then

$$n(r, a) \geq kn(\mu r^k, f) - mn(r, f), \tag{5.18}$$

where $\mu = |c_k| - \delta$, $\delta > 0$.

Proof. Obviously,

$$n(r, f\circ p) \leq n(r, q \circ f) + n(r, a).$$

By (1.1)

$$kn(\mu r^k, f) \le n(r, f \circ p) \le n(r, q \circ f) + n(r, a) \le mn(r, f) + n(r, a),$$

and the assertion follows. □

Theorem 5.3.2. *Let f be a transcendental meromorphic solution of the equation*

$$f(p(z)) = a(z)q(f(z)), \tag{5.19}$$

where a is a meromorphic function, p is a polynomial, $\deg p = k \ge 2$, and q is a polynomial, $\deg q = m$. If a has a finite number of poles and $k > m$, then f has a finite number of poles and if $k = m$, then for any given $\varepsilon > 0$,

$$n(r, f) = O((\log r)^\varepsilon).$$

Proof. Assume that a has exactly N poles and that $k > m$. By (5.18)

$$N = n(r, a) \ge kn(\mu r^k, f) - mn(r, f) \ge (k - m)n(r, f),$$

so $n(r, f)$ is bounded for all r. Assume next that $k = m$. Then from (5.18) we have

$$n(\mu r^k, f) \le n(r, f) + \frac{N}{k}.$$

By Lemma 1.3.2,

$$n(r, f) = O((\log r)^\varepsilon).$$

□

6. The equation $f(p(z)) = a_0 + a_1 f + \cdots + a_{m-2} f^{m-2} + f^m$

In this section we assume that f is a transcendental meromorphic solution of the equation

$$f(p(z)) = a_0(z) + a_1(z)f(z) + \cdots + a_{m-2}(z)f(z)^{m-2} + f(z)^m, \tag{6.1}$$

where $m \ge 2$, the coefficients a_i are small meromorphic functions with respect to f not all identically zero, and p is a polynomial with $\deg p \ge 2$. Moreover, we assume that

$$\overline{N}\left(r, \frac{1}{f \circ p}\right) = S(r, f). \tag{6.2}$$

We obtain a connection between the solutions of the equation (6.1) and the solutions of a certain Riccati differential equation.

6.1. The Clunie lemma

We start with some terminology. Suppose that f is a transcendental meromorphic function and that the coefficients a_i are small meromorphic functions with respect to f. If $j_0, j_1, \dots, j_k$ are nonnegative integers,

$$M[f] = f^{j_0}(f')^{j_1} \dots (f^{(k)})^{j_k}$$

is called a *differential monomial* in f of *degree* $\gamma_M = j_0 + \cdots + j_k$. If $M_1, \dots, M_p$ are differential monomials and if $a_1, \dots, a_p$ are meromorphic functions ($a_i \not\equiv 0$), then

$$P[f] = \sum_{i=1}^{p} a_i(z) M_i[f]$$

is called a *differential polynomial* in f, and the *degree* γ_P is defined by

$$\gamma_P = \max \gamma_{M_i}$$

If P is a differential polynomial, its derivative P',

$$(P[f])' = \frac{d}{dz} P[f],$$

is also a differential polynomial and $\gamma_P = \gamma_{P'}$. The next lemma is due to Clunie [6].

Lemma 6.1.1. *Let Q and Q^* be differential polynomials in f having coefficients a_i and a_i^* respectively. Suppose that $m(r, a_i) = S(r, f)$ and $m(r, a_i^*) = S(r, f)$. If $\gamma_Q \leq n$ and*

$$f^n Q^*[f] \equiv Q[f],$$

then

$$m(r, Q^*[f]) = S(r, f).$$

Remark. The original Clunie's lemma was proved under the assumptions $T(r, a_i) = S(r, f)$ and $T(r, a_i^*) = S(r, f)$. However, it is easy to see that the proof also works with the weaker assumptions stated above. In particular, some of the coefficients may be of the form $\frac{f'}{f}$.

6.2. The connection to the Riccati differential equation

In this subsection we are going to prove that if f is a transcendental meromorphic solution of the equation

$$f(p(z)) = a_0(z) + a_1(z) f(z) + \cdots + a_{m-2}(z) f(z)^{m-2} + f(z)^m$$

under the assumption (6.2), it has to be a solution of a certain Riccati differential equation.

Lemma 6.2.1. *Let f be a transcendental meromorphic solution of the equation*

$$f(p(z)) = a_0(z) + a_1(z)f(z) + \cdots + a_{m-2}(z)f(z)^{m-2} + f(z)^m, \tag{6.3}$$

where $m \geq 2$, the coefficients a_i are small meromorphic functions with respect to f not all identically zero, and p is a polynomial with $\deg p \geq 2$. Assume that

$$\overline{N}\left(r, \frac{1}{f \circ p}\right) = S(r, f).$$

Then there exists a meromorphic function $a \not\equiv 0$ such that $T(r, a) = S(r, f)$ and

$$f'(z) = a(z) + \frac{p'(z)}{m} \frac{f'(p(z))}{f(p(z))} f(z).$$

Proof. The differential polynomial

$$P[f] = a_0 + a_1 f + \cdots + a_{m-2} f^{m-2}$$

is not identically zero by [15]. We differentiate the equation

$$f(p(z)) = f^m + P[f]$$

and get

$$f^{m-1}\widehat{Q}[f] = Q[f], \tag{6.4}$$

where

$$\widehat{Q}[f] = mf'(z) - p'(z)\frac{f'(p(z))}{f(p(z))} f(z)$$

and

$$Q[f] = p'(z)\frac{f'(p(z))}{f(p(z))} P[f] - (P[f])'.$$

Since $m\left(r, \frac{(f \circ p)'}{f \circ p}\right) = S(r, f \circ p) = S(r, f^m + P[f]) = S(r, f)$ and $\gamma_Q \leq m - 2$, we can apply the Clunie lemma. We define

$$a(z) = f'(z) - \frac{p'(z)}{m} \frac{f'(p(z))}{f(p(z))} f(z). \tag{6.5}$$

Hence

$$m(r, a) = S(r, f).$$

If we assume that the function a is identically zero we have

$$f'(z) - \frac{p'(z)}{m} \frac{f'(p(z))}{f(p(z))} f(z) \equiv 0. \tag{6.6}$$

We write equation (6.6) in the form

$$\frac{f'(p(z))}{f(p(z))} = \frac{m}{p'(z)}\frac{f'(z)}{f(z)}.$$

By integrating we see that there exists a non-zero constant C such that

$$f(p(z)) = Cf(z)^m,$$

therefore, by (6.3)

$$(1-C)f^m + a_{m-2}f^{m-2} + \cdots + a_1 f + a_0 \equiv 0.$$

Since the functions a_i are small with respect to f, by [15], we have a contradiction. Therefore, we may assume that a does not vanish identically. Let z_0 be a pole of order μ of a. If f is regular at $z = z_0$, $f \circ p$ must have a pole or zero and $\mu = 1$. If f has a pole at $z = z_0$ of order q and if the coefficients a_i have poles of order not greater than l, then it follows from (6.4) that

$$(m-1)q + \mu \leq 1 + (m-2)q + l$$

or

$$\mu \leq 1 - q + l \leq l.$$

Thus,

$$N(r,a) \leq \overline{N}\left(r, \frac{1}{f \circ p}\right) + \sum_{i=0}^{m-2} N(r, a_i) = S(r,f)$$

by hypothesis. □

Lemma 6.2.2. *Suppose that f is a transcendental meromorphic solution of the equation*

$$f(p(z)) = a_0(z) + a_1(z)f(z) + \cdots + a_{m-2}(z)f(z)^{m-2} + f(z)^m,$$

where $m \geq 2$, the coefficients a_i are small meromorphic functions with respect to f not all identically zero, and p is a polynomial with $\deg p \geq 2$. Assume that

$$\overline{N}\left(r, \frac{1}{f \circ p}\right) = S(r,f).$$

Then

$$\begin{aligned} &(a) \quad m(r,f) = S(r,f),\\ &(b) \quad m\left(r, \tfrac{1}{f}\right) = S(r,f),\\ &(c) \quad N_1(r,f) = S(r,f). \end{aligned}$$

Proof. In the following we use the same notations as in the proof of Lemma 6.2.1.

(a) Dividing (6.4) by $\widehat{Q}[f] = ma$, we get

$$f^{m-2}f = \frac{1}{ma}Q[f]. \tag{6.7}$$

Applying the Clunie lemma to the equation (6.7) where $Q^*[f] = f$, the assertion follows.

(b) We divide the equation (6.5) by af to get

$$\begin{aligned} m\left(r, \frac{1}{f}\right) &= m\left(r, \frac{1}{a}\left(\frac{f'}{f} - \frac{p'(z)}{m}\frac{f'(p(z))}{f(p(z))}\right)\right) \\ &\le T(r,a) + m\left(r, \frac{f'}{f}\right) + m(r,p') + m\left(r, \frac{f'(p(z))}{f(p(z))}\right) + O(1) \\ &= S(r,f) \end{aligned}$$

by the lemma of the logarithmic derivative.

(c) Let z_0 be a pole of f of order $q \ge 2$ and let the coefficients a_i have poles at z_0 of order not greater than l. At z_0 the function a either has a pole of order μ, $(\mu > 0)$, or it vanishes $(-\mu)$ -times, $(\mu \le 0)$. From

$$(m-1)q + \mu \le 1 + (m-2)q + l$$

we deduce that

$$q - 1 \le l - \mu \le l + |\mu|.$$

Therefore, by Lemma 6.2.1,

$$\begin{aligned} N_1(r,f) &\le N\left(r, \frac{1}{a}\right) + \sum_{i=0}^{m-2} N(r, a_i) \\ &\le T(r,a) + \sum_{i=0}^{m-2} N(r,a_i) + O(1) = S(r,f). \end{aligned}$$

The proof is complete. □

Lemma 6.2.3. *Suppose that f is a transcendental meromorphic solution of the equation*

$$f(p(z)) = a_0(z) + a_1(z)f(z) + \cdots + a_{m-2}(z)f(z)^{m-2} + f(z)^m,$$

where $m \geq 2$, the coefficients a_i are small meromorphic functions with respect to f not all identically zero, and p is a polynomial with $\deg p \geq 2$. Assume that

$$\overline{N}\left(r, \frac{1}{f \circ p}\right) = S(r, f).$$

Then there exist meromorphic functions b and $c \not\equiv 0$ such that

$$T(r, b) + T(r, c) = S(r, f)$$

and

$$p'(z)\frac{f'(p(z))}{f(p(z))} = m(b + cf).$$

Proof. Let z_0 be a simple pole of f which is neither a pole of one of the coefficients a_i nor a pole or a zero of a. We divide the equation (6.7) by f^{m-1}, use (6.3) and the properties of logarithmic derivative and by letting $z \to z_0$ we have for the residue R of f,

$$R = -\frac{2a_{m-2}(z_0)}{ma(z_0)}.$$

By Lemma 6.2.2 (a), (c), f has infinitely many simple poles with the required properties, $R(z) \not\equiv 0$ and so $a_{m-2} \not\equiv 0$. Now we define meromorphic functions

$$c = \frac{m}{2}\frac{a(z)}{a_{m-2}(z)}$$

and

$$b = \frac{1}{m}\frac{f'(p(z))}{f(p(z))}p'(z) - cf(z),$$

where $c(z) \not\equiv 0$, $T(r, c) = S(r, f)$ and $m(r, b) = S(r, f)$ by Lemma 6.2.2 (a). By definition of b, $b(z_0) \neq \infty$ at every simple pole of f which is neither a pole of one of the coefficients nor a pole or a zero of a. So,

$$\begin{aligned} N(r, b) &\leq 2N_1(r, f) + \overline{N}\left(r, \frac{1}{f \circ p}\right) + N(r, c) + S(r, f) \\ &= S(r, f) \end{aligned}$$

completing the proof. □

Combining Lemma 6.2.1 with Lemma 6.2.3 we obtain

Theorem 6.2.4. *Suppose that f is a transcendental meromorphic solution of the equation*

$$f(p(z)) = a_0(z) + a_1(z)f(z) + \cdots + a_{m-2}(z)f(z)^{m-2} + f(z)^m,$$

where $m \geq 2$, the coefficients a_i are small meromorphic functions with respect to f not all identically zero, and p is a polynomial with $\deg p \geq 2$. Assume that

$$\overline{N}\left(r, \frac{1}{f \circ p}\right) = S(r, f).$$

Then the function f is also a solution of the Riccati differential equation

$$w' = a + bw + cw^2.$$

7. The equation $f(p(z)) = a_0 + a_1 f + \cdots + a_m f^m$

Here we consider the equation

$$f(p(z)) = a_0(z) + a_1(z)f(z) + \cdots + a_m(z)f(z)^m,$$

where f is a transcendental meromorphic function, p is a polynomial of degree $k \geq 2$, the coefficients a_i are meromorphic, $a_m \not\equiv 0$ and $m > 1$. We study properties of meromorphic solutions while restricting the growth and, respectively, the number of the poles of the coefficients.

Theorem 7.1. *Suppose that f is a transcendental meromorphic solution of the equation*

$$f(p(z)) = a_0(z) + a_1(z)f(z) + \cdots + a_m(z)f(z)^m, \tag{7.1}$$

where p is a polynomial of degree $k \geq 2$, the coefficients a_i are meromorphic, $a_m \not\equiv 0$ and $m > 1$. Suppose further that

$$T(r, a_i) < KT(r^s, f), \qquad i = 0, 1, 2, \ldots, m, \tag{7.2}$$

where K and s are positive constants and r is large enough. If $s < k$, then for given $\varepsilon > 0$

$$T(r, f) = O((\log r)^{\alpha+\varepsilon}),$$

where

$$\alpha = \frac{\log((m+1)K + \frac{m}{s})}{\log \frac{k}{s}}, \text{ if } 1 \leq s < k$$

and

$$\alpha = \frac{\log((m+1)Ks + m)}{\log k}, \text{ if } s < 1 < k.$$

Proof. By (7.1),

$$T(r, f\circ p) = T\left(r, \sum_{i=0}^{m} a_i f^i\right) \le T\left(r, f\sum_{i=1}^{m} a_i f^{i-1}\right) + T(r, a_0) + O(1)$$
$$\le T(r,f) + \sum_{i=1}^{m} T(r, a_i f^{i-1}) + T(r, a_0) + O(1).$$

By induction and (7.2) we have

$$T(r, f\circ p) \le mT(r,f) + \sum_{i=0}^{m} T(r, a_i) + O(1)$$
$$\le (m+\varepsilon)T(r,f) + (m+1)KT(r^s, f) \tag{7.3}$$

Assume that $s \ge 1$. Since $T(r,f)$ is an increasing convex function of $\log r$, $\frac{T(r,f)}{\log r}$ is an increasing function of r. Hence, for any positive constant C and any $t \ge 1$,

$$\frac{T(Cr^t, f)}{T(r,f)} \ge \frac{\log C + t\log r}{\log r} > (1-\varepsilon)t,$$

whence

$$T(r,f) < \frac{1}{(1-\varepsilon)t} T(Cr^t, f) \tag{7.4}$$

for r sufficiently large. So, using (7.4) with $t = s$ and $C = 1$, we have

$$T(r,f) < \frac{1}{(1-\varepsilon)s} T(r^s, f)$$

and deduce that

$$T(r, f\circ p) \le (m+1)KT(r^s, f) + \frac{m+\varepsilon}{(1-\varepsilon)s} T(r^s, f)$$
$$\le \left((m+1)K + \frac{m}{s} + \varepsilon_1\right) T(r^s, f).$$

Then from (1.3)

$$(1-\varepsilon)T(\mu r^k, f) \le \left((m+1)K + \frac{m}{s} + \varepsilon_1\right) T(r^s, f),$$

that is

$$T(\mu r^{\frac{k}{s}}, f) \le \left((m+1)K + \frac{m}{s} + \varepsilon_2\right) T(r, f).$$

Since $k > s$ we have by Lemma 1.3.2 that

$$T(r,f) = O((\log r)^{\alpha_1 + \varepsilon}),$$

where α_1 is as stated in the lemma. Assume next that $s < 1$. Since $\frac{T(r,f)}{\log r}$ is an increasing function of r and $s < 1$, we obtain

$$\frac{T(r,f)}{\log r} \geq \frac{T(r^s,f)}{\log r^s}$$

i.e.

$$\frac{T(r,f)}{T(r^s,f)} \geq \frac{1}{s}. \tag{7.5}$$

From (7.3) and (7.5) we have

$$T(r, f \circ p) \leq ((m+1)Ks + m + \varepsilon)T(r,f)$$

and so from (1.3)

$$T(\mu r^k, f) \leq ((m+1)Ks + m + \varepsilon_3)\, T(r,f).$$

We have by Lemma 1.3.2 that

$$T(r,f) = O((\log r)^{\alpha_2+\varepsilon}),$$

where α_2 is as stated in the lemma. □

The condition $s < k$ in Theorem 7.1 is best possible.

Example 7.2. Given a polynomial $p(z) = c_k z^k + \cdots + c_0$ of degree $k \geq 2$ and a positive integer m, we define the meromorphic functions a_j, $j = 0, \ldots, m$, by

$$a_j(z) = \binom{m}{j} \frac{e^{p(z)}}{(1+e^z)^m}.$$

Then we have

$$\sum_{j=0}^{m} a_j(z) f(z)^j = \frac{e^{p(z)}}{(1+e^z)^m} \sum_{j=0}^{m} \binom{m}{j} f(z)^j = \frac{e^{p(z)}}{(1+e^z)^m}(1+f(z))^m.$$

The functional equation

$$f(p(z)) = \frac{e^{p(z)}}{(1+e^z)^m}(1+f(z))^m$$

has a solution $f(z) = e^z$, which satisfies

$$T(r,f) = \frac{r}{\pi} + O(1).$$

On the other hand, since $k \geq 2$, we have

$$T(r, a_j) = (1+o(1))\frac{|c_k|}{\pi} r^k.$$

Hence Theorem 7.1 does not hold if $s = k$.

In [8] R. Goldstein has proved that if f satisfies the equation

$$f(p(z)) = q(f(z)),$$

where q is a polynomial with $\deg q = \deg p$, then f is a rational function. Here is an example.

Example 7.3. The function $f(z) = \frac{-z^2+z-1}{z}$ is a solution of the equation

$$f(z^2) = 2 + 2f(z) - f(z)^2.$$

Remark. By Example 7.3, it is obvious that for the equation (7.1) we cannot have a result similar to Theorem 2.2.1.

Theorem 7.4. *Let the function f be a transcendental meromorphic solution of the equation*

$$f(p(z)) = a_0(z) + a_1(z)f(z) + \cdots + a_m(z)f(z)^m, \tag{7.6}$$

where the coefficients a_i are meromorphic functions and p is a polynomial of degree $k \geq 2$, be satisfied and suppose that f has infinitely many poles. Suppose further that for some positive constants K and s, and for r large enough, we have

$$n(r, a_i) \leq Kn(r^s, f), \qquad i = 0, 1, \ldots, m. \tag{7.7}$$

If $k > s$, then

$$n(r, f) = O(\log r)^\beta,$$

where A is a constant and

$$\beta = \frac{\log\left(\frac{K+m}{k}\right)}{\log\left(\frac{k}{s_0}\right)},$$

where $s_0 = \max\{s, 1\}$.

Proof. By (7.6), (7.7) and (1.1) we have

$$\begin{aligned} kn(\mu r^k, f) &\leq n(r, f \circ p) \\ &\leq Kn(r^s, f) + mn(r, f) \\ &\leq (K+m)n(r^{s_0}, f). \end{aligned}$$

Since $k > s$ we have by Lemma 1.3.2

$$n(r, f) = O(\log r)^\beta,$$

where β is as stated in the lemma. □

Similarly as earlier the condition $k > s$ in Theorem 7.4 is best possible.

Example 7.5. Given a polynomial $p(z) = c_k z^k + \cdots + c_0$ of degree $k \geq 2$ and a positive integer m, we define the meromorphic functions a_j, $j = 0, \ldots, m$, by

$$a_j(z) = \binom{m}{j} \frac{(1 - e^{-z})^m}{e^{p(z)} - 1}.$$

Then we have

$$\sum_{j=0}^{m} a_j(z) f(z)^j = \frac{(1 - e^{-z})^m}{e^{p(z)} - 1} \sum_{j=0}^{m} \binom{m}{j} f(z)^j = \frac{(1 - e^{-z})^m}{e^{p(z)} - 1} (1 + f(z))^m.$$

The functional equation

$$f(p(z)) = \frac{(1 - e^{-z})^m}{e^{p(z)} - 1} (1 + f(z))^m$$

has a solution $f(z) = \frac{1}{e^z - 1}$, which satisfies

$$n(r, f) = \frac{r}{\pi} + O(1).$$

On the other hand, since $k \geq 2$, we have

$$n(r, a_j) = (1 + o(1)) \frac{|c_k|}{\pi} r^k.$$

References

[1] Ablowitz, M. J., R. Halburd and B. Herbst: On the extension of the Painlevé property to difference equations - Nonlinearity 13, 2000, 889–905.

[2] Barsegian, G. A.: Gamma Lines: On the Geometry of Real and Complex Functions - Taylor & Francis, London, 2002.

[3] Bergweiler, W.: Untersuchungen des Wachstums zusammengesetzter meromorpher Funktionen - Dissertation, Aachen, 1986.

[4] Bergweiler, W., K. Ishizaki, N. Yanagihara: Growth of meromorphic solutions of some functional equations I - Aequationes Math. 63(1-2), 2002, 140–151.

[5] Bergweiler, W., K. Ishizaki, N. Yanagihara: Meromorphic solutions of some functional equations - Methods Appl. Anal. 5(3), 1998, 248–258.

[6] Clunie, J.: On integral and meromorphic functions - J. London Math. Soc. 37, 1962, 17–27.

[7] Clunie, J.: The composition of entire and meromorphic functions- Mathematical Essays Dedicated to A. J. Macintyre, Ohio Univ. Press, 1970, 75–92.

[8] GOLDSTEIN, R.: On certain compositions of functions of a complex variable - Aequationes Math. 4, 1970, 103–126.

[9] GOLDSTEIN, R.: On meromorphic solutions of a functional equation of Ganapathy Iyer - Aequationes Math. 8, 1972, 82–94.

[10] GOLDSTEIN, R.: On meromorphic solutions of certain functional equations - Aequationes Math. 18, 1978, 112–157.

[11] GOLDSTEIN, R.: Some results on factorisation of meromorphic functions - J. London Math. Soc. (2), 4, 1971, 357–364.

[12] GUNDERSEN, G., J. HEITTOKANGAS, I. LAINE, J. RIEPPO, D. YANG: Meromorphic solutions of generalized Schröder equations - Aequationes Math. 63(1-2), 2002, 110–135.

[13] HEITTOKANGAS, J., R. KORHONEN, I. LAINE, J. RIEPPO, K. TOHGE: Complex difference equations of Malmquist type - Comput. Methods Funct. Theory 1, 2001, no.1, 27–39.

[14] HEITTOKANGAS, J., I. LAINE, J. RIEPPO, D. YANG: Meromorphic solutions of some linear functional equations - Aequationes Math. 60, 2000, 148–166.

[15] HELLERSTEIN, S., L. A. RUBEL: Subfields that are algebraically closed in the field of all meromorphic functions - J. Analyse Math. 12, 1964, 105–111.

[16] ISHIZAKI, K.: Hypertranscendency of meromorphic solutions of a linear functional equation - Aequationes Math. 56(3), 1998, 271–283.

[17] LAINE, I.: Nevanlinna Theory and Complex Differential Equations - Walter de Gruyter, Berlin, 1993.

[18] MOHON'KO, A.: The Nevanlinna characteristics of certain meromorphic functions - Teor. Funktsii Funktsional. Anal. i Prilozhen. 14, 1971, 83–87.

[19] MUES, E., N. STEINMETZ: The theorem of Tumura-Clunie for meromorphic functions - J. London Math. Soc. (2), 23, 1981, 113–122.

[20] RITT, J.: Transcendental transcendency of certain functions of Poincaré - Math. Ann. 95, 1925/26, 671–682.

[21] RUBEL, L.: Some research problems about algebraic differential equations II - Illinois Jour. Math. 36, 1992, 659–680.

[22] VALIRON, G.: Sur la dérivée des fonctions algébroides - Bull. Soc. Math. France 59, 1931, 17–39.

[23] WITTICH, H.: Bemerkung zu einer Funktionalgleichungen von H. Poincaré - Arch. Math. 2, 1949/50, 90–95.

University of Joensuu, Department of Mathematics
P.O. Box 111, FIN-80101 Joensuu, Finland
E-mail address: `hsilven@cc.joensuu.fi`